The New

5-Hour

Workweek

How Artificial Intelligence and
Humanoid Robots Will Revolutionize
the Future of Work and Wealth

Title:

The New 5-Hour Workweek: How Artificial Intelligence and Humanoid Robots Will Revolutionize the Future of Work and Wealth

Author:

Lucas Mancini Sandrini

Published by:

Independently published via Amazon Kindle Direct Publishing (KDP)

First Edition

Published in:

November 2024

Disclaimer:

This book is a nonfiction work based on the author's research and analysis. While the author has made every effort to ensure the accuracy of the information contained herein, the opinions and interpretations are solely those of the author. Neither the author nor the KDP platform is responsible for any misuse of the information presented.

To my wife Luiza and my sons Bernardo and Gustavo

Contents

1

Introduction - The New Era of Work

We are on the brink of an unprecedented transformation in the history of humanity. The upcoming technological revolution, driven by artificial intelligence (AI) and humanoid robots, promises to fundamentally redefine the way we work, live, and interact with the world around us.

This new era of work is no longer a distant vision of science fiction; it is an emerging reality that is already beginning to shape entire sectors and influence decisions on a global scale.

Automation is not a new concept. Since the Industrial Revolution, machines have been replacing human labor in various functions, increasing productivity and altering economic dynamics. However, the current wave of technological innovation is entirely different. Modern AI has the ability not only to perform repetitive tasks but also to learn, adapt, and even make complex decisions. When combined with advanced humanoid robots, this technology has the potential to take on roles that, until recently, were considered exclusively human.

AI has been evolving at an accelerated pace in recent years.

Machine learning algorithms and deep neural networks have enabled significant advancements in areas such as speech recognition, computer vision, and natural language processing. Companies worldwide are adopting these technologies to optimize operations, reduce costs, and create new business models.

For example, AI-based virtual assistants like Siri, Alexa, and Google Assistant have become common in households, helping with daily tasks and controlling smart devices. In the financial sector, automated trading algorithms execute transactions in fractions of a second, while fraud detection systems monitor transactions in real time.

While traditional automation has replaced manual and repetitive functions, such as those in factories and automotive assembly lines, humanoid robots represent an evolutionary leap. Designed to physically resemble humans, they can operate in environments designed for people, using tools and interacting with equipment without the need for significant reconfiguration of their surroundings.

Companies like Boston Dynamics and Tesla have demonstrated robots capable of walking, running, jumping, and performing complex tasks in challenging environments. The integration of AI into these robots allows them to learn from their surroundings, recognize objects, and make decisions based on real-time sensory data.

The combination of AI and humanoid robots is poised to impact various sectors, such as:

- **Manufacturing**: Robots capable of complex assembly, quality inspection, and predictive maintenance.
- **Healthcare**: Robotic assistants aiding in surgeries, nursing care, and hospital logistics.

- **Services**: Automated customer service, concierge robots in hotels, and advanced technical support.
- **Construction**: Robots that can build structures, perform controlled demolitions, and even 3D print entire buildings.

This technological transformation promises to increase efficiency, reduce costs, and eliminate human errors. However, it also raises critical questions about the future of employment, wealth redistribution, and the role of humans in society.

Despite the challenges, there are reasons for optimism. Automation has the potential to free humans from dangerous, monotonous, or exhausting jobs, allowing them to focus on more meaningful and fulfilling activities. With the right approach society can enter an era of abundance, where basic needs are met, and individuals have more time to pursue personal passions, lifelong learning, and creative contributions to culture and science.

The new era of work, driven by AI and humanoid robots, is an inevitable reality. The impending changes in the structure of work demand deep reflection on what it means to work and how we can ensure this transition benefits everyone.

This book will explore this journey in detail, from the historical roots of human labor to the future possibilities of a society transformed by technology.

We invite you to join us on this exploration as we navigate the challenges and opportunities defining our path toward a promising future.

2

The History of Human Labor

Since the dawn of humanity, labor has been a central activity in human life, shaping societies, cultures, and economies. The way humans dedicate their time and energy to work has evolved significantly over the centuries, influenced by technological, social, and economic changes. Understanding this evolution is essential to contextualize the transformations that artificial intelligence and humanoid robots will bring in the future.

In hunter-gatherer societies that preceded the Agricultural Revolution, work was directly tied to survival. Small nomadic groups relied on hunting, fishing, and gathering food—activities that required deep knowledge of the natural environment and specific skills. The workday varied depending on the seasons, the availability of resources, and the immediate needs of the group. Interestingly, anthropological studies suggest that these people dedicated an average of four to six hours daily to subsistence activities, leaving them time for rituals, social interaction, and rest. Work was integrated into community life, strengthening social and cultural bonds.

With the domestication of plants and animals around 10,000 years ago, the Agricultural Revolution radically transformed social structures. Agriculture enabled the emergence of permanent settlements, leading to the development of food surpluses and, consequently, the specialization of roles. Farmers worked long hours, from sunrise to sunset, cultivating the land and tending to animals. Land became a valuable resource, giving rise to social hierarchies and economic inequalities. Artisans, merchants, and other roles beyond agriculture emerged, and private property became a central concept in social relations.

In ancient civilizations like Egypt, Mesopotamia, Greece, and Rome, labor was fundamental to the construction of monuments, cities, and impressive infrastructure. Much of this work was carried out by slaves or forced laborers who had no control over their work hours. Society was marked by a sharp class division: a ruling elite, a middle class of merchants and artisans, and a vast working or enslaved class. Despite often inhumane conditions, these workers contributed to significant advances in engineering, architecture, and the development of legal and economic systems. Labor during this period reflected not only economic needs but also power dynamics and domination.

With the fall of the Roman Empire, Europe entered the Middle Ages, characterized by the feudal system. Society was predominantly agrarian, and land remained the primary economic resource. Peasants, or serfs, worked the lands of feudal lords in exchange for protection and the right to cultivate small plots for their subsistence. The workday varied according to the seasons; intense periods of planting and harvesting were followed by times of less agricultural activity. The Church strongly influenced the calendar, with many days dedicated to religious festivals, paradoxically reducing the number of annual workdays. Social hierarchy was rigid, and mobility between classes was virtually nonexistent. Life revolved around the village and obligations to the feudal lord.

The Renaissance, beginning in the 14th century, brought significant cultural and economic changes. The growth of cities and international trade altered the dynamics of work. Specialized artisans organized into guilds, which controlled the quality and prices of products. The great voyages of discovery opened new markets and opportunities, leading to a gradual transition to wage labor and the emergence of a paid working class. Technological innovations, such as the printing press and advances in navigation, drove progress. The bourgeoisie, composed of merchants and bankers, gained prominence and political influence, altering traditional power structures.

In the 18th century, the Industrial Revolution began in England and quickly spread worldwide, profoundly transforming how labor was performed. The introduction of steam engines, new technologies, and production processes drastically changed production methods. Workers, including women and children, faced grueling 12 to 16-hour shifts in often unhealthy and dangerous factory conditions. Industrial discipline imposed strict schedules, intense supervision, and severe penalties for tardiness or mistakes. Urbanization accelerated as large numbers of people migrated from the countryside to cities in search of jobs in emerging industries.

The harsh conditions and exploitation of the labor force during the Industrial Revolution led to growing demands for change. The 19th century saw significant struggles to reduce working hours and improve labor conditions. In the United Kingdom, the Ten Hours Act of 1847 limited the workday for women and children in textile factories. The eight-hour workday movement spread globally, advocating for a division of the day into "eight hours of work, eight hours of leisure, and eight hours of rest." In 1886, a general strike in the United States for the eight-hour workday resulted in the Haymarket events, establishing May 1st as International Workers' Day.

The workday throughout history reflects not only the economic needs of each era but also the prevailing cultural values and power relations. Technology has consistently acted as a driver of change, with innovations leading to shifts in the structure and nature of work. Labor rights advancements were achieved through collective organization promoting social improvements. The pursuit of a healthy balance between work and personal life has been a constant in human history, though challenges and conditions have varied widely.

Understanding the history of human labor is therefore essential to addressing current and future transformations. The evolution of workdays demonstrates that significant changes are possible and that society has the capacity to adapt to new realities.

As we advance into a new era marked by artificial intelligence and humanoid robots, we face similar challenges and opportunities. Just as machines of the Industrial Revolution replaced manual labor, AI has the potential to automate cognitive tasks. Contemporary society must consider how to ensure that the benefits of this technology are widely shared and how to prepare the workforce for the transitions ahead.

There is real potential to further reduce human working hours, leveraging the productivity gains brought by new technologies. This requires a reevaluation of the value of work and a redefinition of what constitutes meaningful employment in an era of technological abundance. Investing in education and developing skills such as creativity, critical thinking, and empathy will be crucial to preparing future generations for a constantly changing world.

3

Technological Revolutions and the Reduction of the Workweek

Throughout history, technological innovation has been a powerful driver of social and economic transformation. Each technological revolution has brought profound changes to how we work, live, and relate to the world.

The Industrial Revolution, which began in the 18th century, marked the first major technological leap that drastically altered the dynamics of work. The invention of the steam engine by James Watt in 1765 was a turning point that enabled the mechanization of production processes previously carried out manually. The steam engine allowed factories to operate continuously, driving mass production and significantly increasing industrial efficiency.

Before the steam engine, production relied primarily on human or animal power, as well as on energy sources like water and wind, which were limited and unpredictable. The new technology made it possible to power machines anywhere and anytime, regardless of weather conditions. This led to the rapid expansion of industries such as textiles, metallurgy, and transportation.

However, this first phase of industrialization did not immediately result in reduced working hours. On the contrary, workers often faced long hours in difficult conditions. Mechanization increased the demand for labor in factories, where men, women, and children endured grueling workdays.

Despite this, mechanization laid the groundwork for future reductions in workloads. As productive efficiency increased, it became evident that fewer workers were needed to produce the same amount of goods. This surplus of labor initiated debates about the need to regulate working hours.

The second great wave of technological innovation came with the discovery and application of electricity in the late 19th and early 20th centuries. Electricity revolutionized not only industry but also daily life. Inventors like Thomas Edison, Nikola Tesla, and others contributed to the development of electrical systems that illuminated cities, powered machines, and transformed communication.

The electrification of factories allowed for unprecedented flexibility in organizing work. Electric machines were safer, more efficient, and easier to operate than steam-powered ones. Moreover, electricity enabled the creation of new products and services, expanding markets and generating jobs in emerging sectors. The automation of repetitive and dangerous tasks reduced the need for heavy manual labor, allowing workers to focus on more skilled and less exhausting roles.

In the domestic sphere, electricity introduced appliances that revolutionized household chores. Devices such as refrigerators, washing machines, and vacuum cleaners lightened the burden of housework, especially for women, who traditionally bore most of these responsibilities. This freed up time for education, paid work, and leisure activities, contributing to significant social changes, such as women entering the workforce.

The third technological revolution, which began in the latter half of the 20th century, was driven by computing and digitalization. The development of electronic computers, initially large and expensive, quickly evolved into accessible and portable personal devices. Moore's Law, observed by Intel co-founder Gordon Moore, predicted that the number of transistors on an integrated circuit would double approximately every two years, exponentially increasing the computing power of devices.

The introduction of computers into the workplace radically transformed productivity and the nature of work. Processes previously performed manually, such as complex calculations, data management, and communication, became automated. Specialized software increased efficiency in sectors like finance, engineering, medicine, and education. The internet, initially a military and academic network, expanded to commercial and domestic use, connecting the world in unprecedented ways.

Computing not only boosted productivity but also enabled flexible work arrangements. Remote work became viable, allowing professionals to perform their duties from anywhere with internet access. This offered a better work-life balance, reduced commuting times, and provided greater schedule flexibility.

The reduction of human workloads as a result of technological revolutions was neither a linear process nor free of challenges. Each technological advance brought benefits and difficulties that society had to address. However, it is undeniable that technology has allowed for more production with less human effort, raising living standards and providing more free time for an increasing share of the population.

History shows that while technology may initially displace jobs, it also creates new opportunities. Agricultural mechanization, for instance, drastically reduced the need for rural labor but freed workers to enter other growing economic sectors. Similarly, industrial automation eliminated certain manufacturing jobs but created demand for technicians, engineers, and IT professionals.

The current technological revolution, driven by artificial intelligence and advanced robotics, has the potential to continue this trend. However, the speed and scale of the changes present unprecedented challenges. Machines' ability to perform complex cognitive tasks raises questions about the future of human work and how society should adapt to these transformations.

As we approach a new technological era, it is crucial to learn from the past and prepare for a future where technology evolves much more rapidly. With careful planning and investment in human capital, we can ensure that upcoming technological revolutions promote not only economic efficiency but also human and social development, leading us toward a society where work is more meaningful and less mandatory—ultimately ensuring that quality of life is widely shared.

4

Industrialization and the Lowering of Goods' Costs

The transition to the industrial era represented one of the most profound changes in human history. Industrialization not only revolutionized production methods but also transformed society in multiple ways. One of the most significant consequences of this period was the lowering of consumer goods' costs, a direct result of mass production.

In the late 18th and early 19th centuries, the Industrial Revolution began gaining momentum in England, later spreading across continental Europe and North America. This period was marked by a series of technological and organizational innovations that transformed artisanal production into mechanized manufacturing. The introduction of steam engines, mechanical looms, and other inventions allowed products to be manufactured on a large scale, faster and more efficiently than ever before.

Before industrialization, the production of goods was predominantly artisanal. Skilled craftsmen created products individually or in small quantities, making goods expensive and inaccessible to most of the population. Production was slow, limited by manual skill and the available time of each artisan. Additionally, the quality and characteristics of the products could vary significantly, as each item was unique.

Industrialization brought about the factory system, where machines operated by workers produced large quantities of standardized products. Mechanization allowed tasks previously performed manually to be executed by machines with greater speed and precision. This significantly reduced production time and associated costs, enabling products to be sold at much lower prices.

One of the most emblematic examples of the impact of mass production was the textile industry. With the invention of the spinning machine and mechanical loom, fabric production increased exponentially. Cotton, once a luxury available only to the wealthy, became accessible to the masses. Quality clothing and fabrics began to be consumed by people across various social classes, improving comfort and hygiene.

Another crucial advancement was the introduction of the assembly line, popularized by Henry Ford in the early 20th century. By applying mass production principles to automobile manufacturing, Ford drastically reduced the production cost of the Model T. The automobile, once a luxury item, became accessible to the American middle class. This not only transformed the automotive industry but also had profound impacts on mobility, urbanization, and 20th-century culture.

Mass production led to the standardization of products. While this meant less customization, it also ensured that consumers could rely on the consistency and quality of the goods they purchased. Products such as shoes, household utensils, tools, and appliances became widely available at prices most people could afford. This elevated living standards, allowing more people to access comforts that were previously unimaginable.

The lowering of goods' costs had profound economic and social impacts. It increased people's purchasing power, stimulating consumption and driving economic growth. Additionally, the accessibility of goods contributed to improved health and well-being. Products like medicines and medical equipment became more available, leading to increased life expectancy and reduced disease rates. Education was also impacted, as the mass production of books and educational materials made them more accessible, promoting literacy and knowledge.

Industrialization laid the foundation for continuous technological development. Competition among companies encouraged innovation, leading to the creation of new products and technologies. The efficiency of mass production allowed society to allocate resources to other areas, such as scientific research, education, and infrastructure, promoting progress on multiple fronts.

Globalization, fueled by advances in transportation and communication, further expanded the reach of mass production. Goods could be produced in one country and sold worldwide, integrating economies and cultures. This increased the diversity of products available to consumers and fostered cultural exchange.

The democratization of consumption altered social dynamics. The distinction between social classes became less pronounced in terms of access to material goods. While economic disparities persisted, the availability of basic products and even some luxuries, like televisions and smartphones, to the majority of the population contributed to a sense of inclusion and participation in consumer society.

In the current context, industrialization and the lowering of goods' costs continue to evolve with the introduction of new technologies such as advanced automation, artificial intelligence, and 3D printing. These innovations have the potential to further reduce production costs and customize mass-produced goods, catering to individual consumer preferences without sacrificing efficiency.

Industrialization and mass production have had a profound impact on the accessibility and cost of goods, transforming society in significant ways. The lowering of goods' costs allowed more people to access products and services that improved quality of life, promoted economic development, and stimulated innovation. While it brought challenges, industrialization undeniably paved the way for a more connected and prosperous society.

5

The Second Half of the 20th Century: The Dawn of the Digital Era

The second half of the 20th century marked the beginning of a new era in human history: the Digital Age. This phase was characterized by unprecedented technological advancements that radically transformed how we live, work, and relate to one another. Automation, made possible by the development of computers and digital technologies, emerged as one of the main drivers of this transformation, redefining production processes and propelling productivity to levels never before imagined.

The technological evolution that culminated in the Digital Age had its roots in the development of the first electronic computers during World War II. Machines like the ENIAC (Electronic Numerical Integrator and Computer) demonstrated the potential of computers to process large volumes of data and perform complex calculations at unimaginable speeds for the time. Initially used for military and scientific purposes, computers gradually found applications in civilian sectors, driving the automation of administrative and industrial tasks.

The 1960s saw the miniaturization of electronic components, thanks to the invention of the transistor and, later, integrated circuits. These advances enabled the creation of smaller, faster, and more affordable computers. Companies like IBM led the commercialization of computers for business use, facilitating the automation of processes such as accounting, inventory management, and data processing. The introduction of these systems resulted in significant efficiency gains, reducing human errors and accelerating critical operations.

Automation was not limited to office environments. In industries, the adoption of computer numerical control (CNC) systems revolutionized manufacturing. Computer-controlled machine tools could produce parts with millimeter precision, operate 24 hours a day, and quickly adjust to different specifications. This enabled flexible mass production, combining the efficiency of large-scale production with the ability to customize.

The 1970s marked the advent of microprocessors, which fueled the creation of personal computers. Companies like Apple and Microsoft emerged, making computing accessible to individuals and small businesses. The personal computer (PC) became an indispensable tool, democratizing access to information and automation technologies. Software programs for word processing, spreadsheets, and databases transformed individual and organizational productivity.

In the 1980s, automation advanced with the introduction of management information systems (MIS) and enterprise resource planning (ERP) systems. These tools integrated various organizational functions, such as finance, human resources, and logistics, providing a comprehensive view of business processes. The ability to analyze real-time data improved decision-making and optimized operations, reducing waste and increasing efficiency.

The emergence of the internet, initially a military and academic communication network, evolved into a global platform for communication and commerce in the 1990s. The World Wide Web, developed by Tim Berners-Lee, facilitated access to information and connected people and organizations on a global scale. The internet became a catalyst for automating business processes, from financial transactions to supply chains.

E-commerce emerged as a new frontier, with companies like Amazon and eBay revolutionizing how products and services were bought and sold. The automation of sales, customer service, and logistics processes allowed businesses to operate globally at reduced costs. The digitization of products, such as music and movies, transformed entire industries, eliminating the need for physical media and enabling instant distribution.

In the industrial environment, automation advanced with the introduction of industrial robots. Robotic equipment, initially used for dangerous or repetitive tasks, became more sophisticated, capable of performing complex operations with high precision. Collaborative robotics, or "cobots," enabled humans and robots to work side by side, combining human flexibility with the strength and precision of machines.

Artificial intelligence (AI) began to gain prominence as a tool to enhance automation. Machine learning algorithms and neural networks enabled computational systems to learn from data, identify patterns, and make predictions. AI applications were implemented in areas such as fraud detection, medical diagnosis, market analysis, and product personalization. The ability of intelligent systems to adapt and improve over time expanded the potential of automation beyond pre-programmed tasks.

Robotic process automation (RPA) emerged as a solution for automating repetitive administrative tasks. Software bots could interact with existing systems, performing operations such as data entry, transaction processing, and responding to queries. This freed employees to focus on higher-value activities such as analysis, strategy, and innovation.

The impact of automation on productivity was significant. Companies that adopted automated technologies experienced cost reductions, improvements in product and service quality, and greater responsiveness to market demands. Operational efficiency became a competitive advantage, accelerating the adoption of digital technologies.

Globally, automation contributed to shifts in international competitiveness. Countries that invested in technology and innovation gained economic advantages, while those that lagged behind faced challenges in maintaining productivity and growth. Technological disparities widened economic gaps between nations, highlighting the importance of policies that promote digital inclusion.

Undeniably, the Digital Age and automation profoundly transformed productivity and the global economic structure. The ability to automate processes expanded human possibilities, allowing us to focus on more complex and creative activities.

History demonstrates that technology alone does not determine social and economic outcomes. The choices made by society in terms of policies, education, and values are crucial in shaping technology's impact positively. Automation has the potential to create an era of abundance and shared prosperity, but this depends on how we address challenges and seize opportunities.

As we move into the future, it is essential to continue exploring ways to integrate technology in a manner that promotes human well-being. The Digital Age is a step in an ongoing journey of transformation, and it is up to us to shape its course to create a fairer, more productive, and sustainable society.

6

Artificial Intelligence: The Next Leap

The exploration of artificial intelligence (AI) as a means to replace complex cognitive tasks is one of the primary drivers of the contemporary technological revolution. AI has transitioned from being merely a theoretical concept or a science fiction element to becoming a real and transformative force in modern society. Its influence is reshaping entire industries, redefining jobs, and creating new economic opportunities.

But what exactly is Artificial Intelligence? Simply put, AI is a branch of computer science dedicated to creating systems capable of performing tasks that typically require human intelligence. This includes abilities such as learning, reasoning, perception, language comprehension, and problem-solving. AI aims to emulate human cognitive processes, enabling machines to not only execute pre-programmed instructions but also learn from experience and adapt to new situations.

In recent years, significant advancements in areas such as machine learning and deep learning have propelled AI to unprecedented levels of capability and efficiency. Sophisticated algorithms can now analyze vast amounts of data, identify complex patterns, and make informed decisions with speed and precision that surpass human abilities. This has paved the way for revolutionary applications in various sectors, from healthcare and education to finance and transportation.

AI's ability to replace complex cognitive tasks has profound implications for the world of work. Professions once considered irreplaceable by automation are now being transformed. Financial analysts, lawyers, doctors, and even artists are seeing parts of their activities taken over by AI systems. For instance, AI-assisted medical diagnostic programs can detect diseases at early stages with greater accuracy, while legal software can analyze legal documents in a fraction of the time it would take a human.

This technological transformation is not just about substitution but also about enhancing human capabilities. AI can handle tedious and repetitive tasks, allowing professionals to focus on more creative and strategic aspects of their work. In a scenario where the workweek is reduced to five hours, AI is the ally that makes this model sustainable by automating processes and boosting productivity.

The harmonious integration of AI into society depends on balancing innovation with responsibility, as ethical issues related to autonomous decision-making by machines, data privacy, and algorithmic bias require careful attention. Proper regulations and ethical guidelines are essential to ensure that AI development benefits everyone and not just a privileged few. Transparency in algorithms and accountability for decisions made by AI systems are crucial aspects of this process.

In the business environment, AI is redefining business models. Companies adopting AI technologies gain significant competitive advantages by optimizing operations, improving customer experiences, and exploring new markets. AI-focused startups are emerging at a rapid pace, attracting substantial investments and driving innovation in fields such as robotics, natural language processing, and computer vision.

For individuals, AI offers the opportunity to better balance personal and professional life. With routine tasks automated, people can dedicate more time to activities that promote well-being, continuous learning, and personal development. The promise of a reduced workweek becomes viable when AI technologies are used to maximize efficiency and minimize time waste.

Ultimately, Artificial Intelligence represents the next leap in humanity's technological evolution. Its ability to take on complex cognitive tasks not only changes how we work but also how we perceive our role in the world. The key to fully leveraging this potential lies in adaptation, education, and the pursuit of solutions that promote shared prosperity.

As we move toward a future where AI is ubiquitous, it is crucial to maintain a focus on the human values that define our society. Empathy, creativity, ethics, and social awareness are attributes that AI cannot yet fully replicate. Therefore, the collaboration between humans and intelligent machines will be the foundation on which we build a better world—where a five-hour workweek is not just an ideal but an achievable reality.

7

Humanoid Robots in the Workplace

The arrival of humanoid robots in the workplace represents one of the most significant evolutions of the current technological era. The idea of machines capable of performing human tasks is not new; however, recent advances in robotics and artificial intelligence have turned this vision into a tangible reality. Advanced robots are gradually taking on human roles across various sectors, reshaping the way we conceive work and redefining humanity's role in modern society.

Humanoid robots are machines designed to resemble the human body in form and function. This resemblance is not merely aesthetic; it enables robots to operate in environments designed for humans, using tools and interacting with equipment without the need for adaptations. The anthropomorphization of these robots facilitates their integration into existing spaces, making the transition smoother and more efficient.

In the industrial sector, for instance, humanoid robots are being introduced to perform tasks requiring manual dexterity and real-time decision-making. On assembly lines, they can execute complex operations once exclusive to human workers, such as assembling delicate components or conducting quality inspections. Automotive and electronics companies are heavily investing in this technology to enhance production precision and speed, reduce errors, and minimize operational costs.

In healthcare, humanoid robots are set to revolutionize patient care. They may assist in surgeries, offering precision beyond human capabilities, or act as caregivers in clinics and hospitals, performing tasks such as administering medication, monitoring vital signs, and even providing companionship to isolated patients.

The service sector is also benefiting from this technology. In hotels and restaurants, humanoid robots will soon serve as receptionists, waitstaff, and attendants, interacting with customers naturally and efficiently. They will be able to understand and respond to requests in multiple languages, improving the customer experience and operating without rest, increasing productivity. The Henn-na Hotel in Japan is a pioneering example, where humanoid robots already perform the majority of operational functions.

In education, these robots will soon be employed as teaching assistants, offering personalized support to students. They can adapt teaching methods to individual needs, assist in interactive learning activities, and even help children with special needs. Interaction with humanoid robots can make the educational process more engaging, stimulating student interest in science and technology.

In the security and defense sectors, humanoid robots are being developed to carry out high-risk missions, such as bomb disposal, disaster rescue operations, and patrolling dangerous areas. Their ability to operate in hostile environments without endangering human lives is one of the main benefits of this application. Companies like Boston Dynamics have made significant strides, creating robots capable of navigating complex terrains and performing critical tasks autonomously.

The integration of humanoid robots in the workplace raises important questions about the future of employment. Legitimate concerns exist about the replacement of human workers and the potential rise in unemployment. However, it is crucial to understand that automating routine and repetitive tasks can free humans to focus on activities requiring creativity, critical thinking, and emotional intelligence—skills that robots cannot yet fully replicate.

Moreover, the development and maintenance of humanoid robots create new employment opportunities in fields such as engineering, programming, interface design, and technological ethics. The economy can benefit from increased efficiency and productivity, driving economic growth that, if well-managed, could result in an overall improvement in quality of life.

Social acceptance of humanoid robots is also a determining factor for their success. Public perception of the reliability and intentions of these machines influences how quickly they will be adopted on a large scale. Educational campaigns and transparency in development processes can help build trust and understanding of robots' roles in the future of work.

Concrete examples of humanoid robots making a difference include Tesla's Optimus, introduced under Elon Musk's leadership. This robot is designed to perform repetitive and dangerous tasks, with plans for commercialization in the near future. Simultaneously, OpenAI has partnered with the robotics startup Figure to integrate advanced AI systems, such as GPT, into humanoid robots. This collaboration aims to improve the interaction and adaptability of these machines, enabling them to adjust to diverse contexts and situations more efficiently and intuitively.

In sectors like agriculture, humanoid robots can perform harvesting, monitor crops, and optimize natural resource usage. In logistics, they can operate warehouses, manage inventories, and execute deliveries with efficiency and precision. Even in space exploration, humanoid robots are being considered for missions requiring human-like capabilities in environments where direct human presence is risky or unfeasible.

Collaboration between humans and robots is a promising area. Rather than fully replacing workers, robots can act as assistants, enhancing human capabilities and enabling tasks to be completed more effectively. This synergy could lead to innovations once thought unimaginable, driving progress across various fields.

In terms of economic impact, adopting humanoid robots can lead to significant reductions in operational costs, increased production, and the opening of new markets. Companies leading this transformation will gain substantial competitive advantages but also bear the responsibility of driving this change ethically and sustainably.

Culture and the arts will also be influenced. Robots capable of creating music, art, or literature challenge our conceptions of creativity and originality. This intersection between technology and human expression opens fascinating debates about what it means to be creative and how we value artistic production.

The future of work with humanoid robots is not mere science fiction but an emerging and inevitable reality that demands our attention and action. By embracing this change responsibly and conscientiously, we can create a world where humans and machines collaborate to achieve new heights of innovation and prosperity, making a shorter workweek and a superior quality of life a reality for all.

8

The 5-Hour Workweek Model

The vision of a future where the workweek is drastically reduced is not merely a distant utopia but a realistic possibility emerging from the convergence of unprecedented technological advancements, particularly in artificial intelligence and humanoid robotics.

As previously noted, the 20th century saw the adoption of the 40-hour workweek as the standard in many countries. However, since then, reductions in working hours have stagnated, even with technological advancements that have significantly increased productivity. According to economic studies, labor productivity in developed countries has grown exponentially in recent decades. For example, data from the Organization for Economic Cooperation and Development (OECD) indicates that productivity per hour worked in the United States has increased by approximately 250% since 1950. Yet, this increase has not translated into a proportional reduction in working hours.

The massive introduction of artificial intelligence and humanoid robots into the labor market has the potential to break this paradigm. With machines capable of performing cognitive and physical tasks more efficiently than humans, the need for traditional labor can be significantly reduced.

Consider some calculations to illustrate this point.

Suppose a factory employs 1,000 human workers, each working 40 hours a week, totaling 40,000 weekly work hours. With the implementation of humanoid robots capable of working 24 hours a day, 7 days a week, without fatigue or the need for rest, the same factory could operate with far fewer robotic units. If a robot can work continuously, it provides 168 hours of work per week (24 hours × 7 days). To match the 40,000 weekly hours, approximately 238 robots would be needed (40,000 hours / 168 hours per robot ≈ 238). This represents a significant reduction in the need for human workers.

Additionally, robots are not subject to human errors, breaks, vacations, or sick leave, further increasing operational efficiency. With proper maintenance and software updates, they can continually improve their performance. Although the initial costs of acquisition and implementation may be high, the long-term economic benefits are substantial. Studies by the McKinsey Global Institute estimate that automation could boost global productivity by up to 1.4% annually.

Future projections suggest that, if current trends continue, we could see a significant reduction in working hours in the coming decades. Economist John Maynard Keynes, in his 1930 essay "Economic Possibilities for Our Grandchildren," predicted that by 2030, the workweek would be reduced to 15 hours due to technological advances. While his prediction was optimistic for its time, the possibility of an even shorter workweek is not out of reach, particularly with the acceleration of AI and robotics.

Imagine a scenario where most productive tasks are carried

out by intelligent machines. The production of goods and services would become abundant, and associated costs would be drastically reduced. This could lead to a new economic model where income and wealth distribution are reorganized to ensure the benefits of automation are shared across society, not just concentrated among the corporations owning the technology.

A 5-hour workweek model would allow individuals more time for personal, family, educational, and leisure activities. Psychological studies indicate that a better quality of life is associated with a healthy balance between work and personal life. With more free time, people could focus on personal development, volunteering, and activities that promote social well-being.

However, this vision faces significant challenges. Implementing an economic model that supports such a reduced workweek requires profound structural changes. Public policies aimed at income redistribution, such as universal basic income, could be considered. Moreover, issues related to intellectual property, where all AI-generated content could become the collective property of humanity, must be carefully addressed.

Historically, the introduction of disruptive technologies has always sparked concerns about technological unemployment. During the Industrial Revolution, Luddites destroyed machines that threatened their jobs. Over time, however, new industries and job opportunities emerged, and society adapted. The difference now is the scale and speed at which AI and robots are advancing. Automation is no longer limited to manual tasks but also encompasses complex cognitive functions.

Mathematically, if we consider Moore's Law, which observes the exponential growth of computational capacity, we can infer that AI systems will continue to improve at an accelerated rate. This means that tasks currently considered complex for machines may soon be automated. For example, deep learning algorithms are becoming proficient in areas like medical diagnosis, legal analysis, and even composing music and texts.

It is projected that by 2030, up to 800 million jobs worldwide could be automated, according to a McKinsey report. This represents about one-fifth of the global workforce.

Consider, for instance, a shopping center years ago where numerous employees worked as parking lot attendants. Each exit was manually managed, with staff verifying payments and raising barriers. Today, this operation is almost entirely automated, requiring only a couple of employees to oversee machine functionality. This scenario illustrates a classic example of replacing dozens of human workers with machines, resulting in a significant productivity increase—estimated to be at least tenfold.

In terms of productivity, the full automation of certain industries could lead to a dramatic increase in global output. If productivity doubles every 10 years with the implementation of advanced AI and robotics, within 30 years, we could see an eightfold increase in current productivity. This means the same amount of goods and services could be produced with only a fraction of the human labor currently required.

However, for the 5-hour workweek to be viable and sustainable, addressing the distribution of productivity gains is crucial. If the wealth generated by automation remains concentrated in the hands of a few, social and economic disparities will increase. Progressive tax policies, investments in social infrastructure, and wealth-sharing mechanisms will be essential to avoid extreme inequalities.

The integration of humanoid robots into the labor market also raises cultural and psychological questions. Daily interactions with machines simulating human behavior could affect how we relate to one another. Developing social and ethical norms to guide these interactions will be necessary to ensure that the presence of robots does not dehumanize social relationships.

In sectors like agriculture, automation could revolutionize food production. Robots capable of monitoring soil, planting, tending, and harvesting crops could increase efficiency and reduce waste. This would not only ensure global food security but also free rural workers to pursue other opportunities and experiences.

In the service sector, AI and humanoid robots can offer personalized services on a massive scale. Virtual assistants are already being used to support customers across various industries. With advancements in artificial emotional intelligence, these assistants will increasingly understand and respond to customer needs more sophisticatedly.

In the transportation sector, autonomous vehicles will replace human drivers, enhancing safety and travel efficiency. Considering there are millions of professional drivers worldwide, automating this sector could free up a significant amount of labor for other activities.

In education, AI can personalize learning for each student, tailoring content to individual needs. This could lead to a better-educated population, equipped to tackle the challenges of a constantly evolving job market.

From an environmental perspective, automation can contribute to more sustainable practices. Robots can optimize natural resource usage, reduce waste, and monitor environmental impacts in real time. AI can model complex scenarios to assist in decision-making that promotes sustainability.

In conclusion, the 5-hour workweek model is an ambitious but achievable vision dependent on several interconnected factors. The key to realizing this vision lies in how we manage the transition to a future dominated by artificial intelligence and robotics. Governments, businesses, and civil society must collaborate to create structures that promote equity, social justice, and collective well-being.

The 5-hour workweek does not have to be a distant dream. With the power of artificial intelligence and humanoid robots, we have the tools to build a future where work is not only a means of sustenance but also an expression of human purpose and creativity. It is up to us to shape this future in a way that reflects our deepest values and collective aspirations.

9

Wealth and Quality of Life in the New Paradigm

The revolution brought about by artificial intelligence and humanoid robots is not only transforming the way we work but also redefining the concepts of wealth and quality of life. In this new paradigm, the distribution of wealth and the increase in leisure and personal development time are direct outcomes of an economy driven by advanced technology.

The massive introduction of AI and humanoid robots into production processes has the potential to elevate economic efficiency to unprecedented levels. When intelligent machines take over the majority of tasks, the production of goods and services can increase exponentially, while associated costs decrease significantly. This creates an abundance of resources that, if well-managed, can raise the overall standard of living.

To illustrate this point, consider a simplified economic model. Suppose a factory produces 1,000 units of a product per day, employing 100 human workers, each earning a daily wage of $100. The daily labor cost is, therefore, $10,000. With the implementation of humanoid robots, which have an operational cost of $20 per unit per day (including maintenance and energy), the factory can produce at least three times more, as robots can work 24 hours a day with significantly lower costs.

If the factory replaces the 100 workers with 100 robots, production could potentially triple to 3,000 units due to the continuous efficiency of robots, which operate without breaks or human errors. The daily operational cost of the robots would be $2,000 (100 robots x $20), a drastic reduction compared to the $10,000 previously spent on wages. This results in a daily savings of $8,000 and a threefold increase in productivity, which could be reinvested into the company, distributed to shareholders, or ideally passed on to consumers in the form of lower prices.

The reduction in production costs has a cascading effect on the economy. Cheaper products increase consumers' purchasing power, enabling them to acquire more goods and services with the same income. Moreover, with the reduced need for human labor in routine tasks, people have more free time to engage in activities that promote personal development and well-being.

It is important to highlight that this model does not depend on forced wealth redistribution but rather on value creation through technological efficiency. Market competition will encourage companies to pass some of the savings on to consumers to remain competitive. This aligns with classical economic principles of supply and demand without requiring redistributive government interventions.

Another area where AI and humanoid robots positively impact is per capita productivity. According to World Bank data, global average productivity (measured by GDP per capita) has gradually increased over the decades. With automation, this growth will accelerate. If per capita productivity increases by 5% annually due to automation, instead of the current 2%, in 20 years, productivity would be approximately 165% higher than today, compared to 49% with the current growth rate.

Mathematically, this can be demonstrated through the compound growth formula:

$$P = P_0 \times (1 + r)^n$$

Where:

- P represents future productivity;
- P_0 is the current productivity;
- r is the annual growth rate (as a decimal);
- n is the number of years.

With a growth rate of 5%:

$$P = P_0 \times (1 + 0,05)^{20} = P_0 \times 2,653$$

With a growth rate of 2%:

$$P = P_0 \times (1 + 0,02)^{20} = P_0 \times 1,4859$$

The difference is significant, indicating that automation can almost double the increase in productivity compared to the current race.

This increase in productivity allows society to produce more with fewer human resources, freeing up time for people to pursue activities that enrich their lives in other ways. With more free time, there is a potential rise in the consumption of services related to leisure, education, culture, and healthcare, stimulating economic sectors that value creativity and human interaction.

For instance, the tourism industry could experience a boom, as more people have the time and resources to travel. The education sector could expand, with individuals seeking lifelong learning either for personal fulfillment or to adapt to changes in the job market. The demand for artistic and cultural activities may increase, fostering creativity and innovation.

From an individual perspective, the reduction in working hours and the increase in free time are associated with improvements in mental and physical health. Psychological studies indicate that stress related to excessive work contributes to various illnesses, including depression, anxiety, and cardiovascular problems. With more time for physical exercise, hobbies, and social relationships, the population could show significant improvements in public health indicators.

Furthermore, the possibility of dedicating time to entrepreneurship increases. Individuals with time and resources can start their own businesses, innovate, and contribute to economic dynamism. This could lead to the emergence of new industries and opportunities, fueling a virtuous cycle of growth and prosperity.

However, for this new paradigm to be sustainable, it is necessary to consider the issues of employability and income. As automation replaces traditional roles, job opportunities must shift toward areas that value uniquely human skills, such as critical thinking, creativity, and emotional intelligence.

A practical example is the information technology sector. With the exponential growth of AI, there is a growing demand for professionals specializing in data science, algorithm development, and cybersecurity. These are fields that require advanced knowledge and adaptability, areas where continuous education is essential.

Economically, the wealth generated by automation can stimulate investments in infrastructure, research, and development. Countries that strategically adopt these technologies can position themselves at the forefront of the global economy, attracting talent and international capital.

Mathematically, if we consider that a country's economy grows at a rate of 3% annually without automation but could grow at 5% with widespread implementation of AI and robotics, the accumulated difference over 30 years is substantial. Using the compound growth formula again:

Without automation:

$$PIB_{30} = PIB_0 \times (1 + 0,03)^{30} = PIB_0 \times 2,427$$

With automation:

$$PIB_{30} = PIB_0 \times (1 + 0,05)^{30} = PIB_0 \times 4,322$$

This means that GDP would be nearly double with the accelerated growth driven by automation.

It is important to note that this economic growth does not rely on the adoption of redistributive policies but rather on wealth creation through efficiency and innovation. Companies that invest in technology can achieve significant returns, but they also face greater competition, which benefits consumers.

In the global context, automation can help reduce inequalities between countries. Developing nations can adopt AI and robotics technologies to boost their economies, improve agricultural, industrial, and service productivity, and provide better opportunities for their populations. This can accelerate economic growth and raise living standards, contributing to international stability.

Companies can play a key role in promoting responsible practices. By investing in local communities, offering education programs, and adopting sustainable business models, they can contribute to balanced development.

Additionally, increased energy efficiency and waste reduction are added benefits of automation. Robots can operate in optimized ways, reducing energy consumption and minimizing environmental impacts. This contributes to sustainability and meets the growing demand for eco-friendly practices.

In summary, wealth and quality of life in the new paradigm are direct outcomes of the intelligent integration of artificial intelligence and humanoid robots into the economy. By increasing productivity, reducing costs, and freeing up time for personal development, these technologies have the potential to positively transform society.

This new paradigm is not a break from established economic principles but a natural evolution driven by technology. By embracing the opportunities offered by AI and robotics, we can achieve levels of wealth and well-being once thought unimaginable, creating a society where free time is valued and human potential is fully realized.

10

Preparing for the Future: Educational and Professional Adaptation to the New Reality

H umanity stands on the brink of an unprecedented transformation, driven by the rise of artificial intelligence and humanoid robots taking over most tasks previously performed by humans. In this context, educational and professional adaptation becomes not just a necessity but a sine qua non condition for thriving in this new era. Preparing for the future requires a profound reevaluation of educational systems, the skills valued in the labor market, and, fundamentally, how we perceive our role in the world.

History offers valuable lessons during periods of significant change. In Ancient Greece, philosophers grappled with existential and practical questions as they sought to understand the nature of knowledge, reality, and ethics. Socrates, Plato, and Aristotle dedicated their lives to developing critical thinking, emphasizing the human ability to reason, question, and seek truth. This philosophical legacy provides a foundation for addressing current challenges, as it emphasizes skills that remain uniquely human, even in the face of the most advanced technologies.

As machines become more proficient in cognitive and physical tasks, education must shift its focus from memorization and information reproduction to the development of critical thinking, creativity, and emotional intelligence. Skills such as complex problem-solving, innovation, and empathy become fundamental. Pedagogy should encourage intellectual curiosity, fostering environments where questioning and exploration are encouraged.

Consider for example, the impact of AI on medicine. Robotic surgeons and advanced diagnostic systems can surpass the precision and speed of human professionals in many areas. However, communication with patients, understanding emotional and cultural nuances, and making ethical decisions remain human domains. Therefore, the training of future doctors must emphasize not only technical knowledge but also interpersonal and ethical skills.

In the professional realm, adaptability is the new currency of value. Traditional careers can become obsolete within a few years, while new professions emerge at a rapid pace. A mindset of continuous learning is essential. Professionals must be willing to reinvent themselves, acquiring new skills and exploring different fields of knowledge. This reflects the Socratic ideal of recognizing one's ignorance as a starting point for wisdom.

Integrating Greek philosophy as a mode of critical thinking in modern education offers tools to navigate the complexity of the contemporary world. Socrates' dialectical method, for instance, stimulates deep questioning and the pursuit of well-founded answers. This method can be applied to the teaching of sciences, technology, and the humanities, promoting a more holistic and critical understanding of subjects.

Moreover, Aristotelian ethics, centered on the pursuit of virtue and the common good, is especially relevant. As AI and humanoid robots take on significant roles in society, ethical questions emerge prominently. How do we program machines to make moral decisions? What is the societal impact of automated actions? Educational training should include ethical and philosophical discussions to prepare individuals capable of addressing these dilemmas.

Practically speaking, governments and educational institutions need to invest in curriculum reform. Subjects that promote transversal skills, such as philosophy, arts, and physical education, should gain prominence alongside STEM (science, technology, engineering, and mathematics). Collaboration across different fields of knowledge is vital to fostering innovation and creativity.

In the professional environment, companies should promote a culture of learning and flexibility. Continuous training programs, opportunities for internal mobility, and encouragement of experimentation can help retain talent and prepare them for change. Business leaders need to value not only technical skills but also soft skills like effective communication, leadership, and empathy.

Artificial intelligence and humanoid robots also raise questions about the meaning of work in human life. With the reduced need for manual and repetitive labor, people have the opportunity to pursue achievements in areas that were previously inaccessible. This could lead to a reevaluation of free time, leisure, and personal development. Epicurean philosophy, which values the pursuit of moderate pleasure and peace of mind, offers insights into how to embrace this new reality.

However, the transition to this future is not without challenges. Inequality could increase if access to quality education and reskilling opportunities is not widely available. Individuals and communities unable to adapt risk being left behind.

Technology itself is neither good nor bad; it is a tool that reflects the values of the society that uses it. Platonic ethics emphasize the importance of a just society, where each individual contributes according to their abilities and receives according to their needs.

When it comes to artificial intelligence, transparency and accountability are fundamental. Developers and engineers must ensure that algorithms are fair, avoiding biases that could harm specific groups. Human oversight is essential to monitor and correct potential ethical deviations. This is an area where philosophical critical thinking can make significant contributions, questioning assumptions and analyzing the implications of technological decisions.

International collaboration also plays a crucial role. The global nature of technology requires cooperation between countries to establish ethical and regulatory standards that protect human interests. Inspired by the cosmopolitan Stoic philosophy, which considers all humans as members of a global community, we can work together to address shared challenges.

Finally, preparing for the future involves a shift in mindset. Instead of fearing replacement by machines, we should embrace the opportunities they provide. Artificial intelligence and humanoid robots free us from the limitations of physical and repetitive work, allowing us to fully explore our intellectual and creative potential. It is an opportunity to redefine what it means to be human in a technologically advanced world.

Education must, therefore, cultivate not only knowledge but also wisdom. As the Greek philosophers taught us, wisdom is the practical application of knowledge to live a good and just life. In this new paradigm, wisdom will guide us to use technology in ways that enrich our lives and promote human progress.

11

Embracing Abundance

Humanity is on the verge of entering an era of unprecedented abundance, driven by artificial intelligence and humanoid robots taking over the majority of tasks previously performed by humans. This new paradigm promises not only to improve efficiency and productivity but also to fundamentally transform the way we live, work, and interact with one another. By embracing this abundance, we can imagine a future where economic and social limitations are overcome, paving the way for a more prosperous and harmonious society.

Advanced artificial intelligence, combined with humanoid robotics, is redefining the foundations of the global economy. With machines capable of learning, adapting, and performing complex tasks with superhuman precision, the production of goods and services theoretically becomes unlimited. Scarcity, which has historically been a determining factor in economics, could gradually be eliminated. This allows us to reimagine fundamental concepts such as ownership, value, and exchange.

One sector poised for radical transformation is the stock market. Traditionally, stock exchanges function as platforms where investors buy and sell shares of companies, basing their decisions on performance analyses, future expectations, and financial strategies. However, with artificial intelligence dominating decision-making, human trading becomes obsolete. High-frequency algorithms already play a significant role in today's markets, but the next evolution could lead to complete automation.

Imagine a scenario where all companies are managed by artificial intelligences optimized to maximize efficiency and innovation. These AIs communicate directly with each other, negotiating resources, forming partnerships, and adjusting strategies in real time. The stock market, as we know it, loses its relevance because investment decisions are no longer based on human speculation but on instantaneous and precise analyses conducted by the AIs themselves. Market fluctuations, often driven by human emotions like fear and greed, are smoothed out or eliminated, resulting in unprecedented economic stability.

In this context, the role of the human investor is profoundly altered. Instead of attempting to outperform the market through investment strategies, individuals can focus on other forms of contribution and personal fulfillment. The wealth generated by automated companies can be distributed more equitably—not through government imposition, but as a natural outcome of an economy where production costs are drastically reduced and efficiency is maximized.

Another significant example is the transformation of the consumer market. Today, price competition is one of the primary drivers of the economy, with companies vying for consumer preference through discounts and promotions. However, with artificial intelligence and humanoid robots reducing production and distribution costs to near-zero levels, the price of products is likely to drop significantly. Additionally, AI systems can instantly analyze all available offers, ensuring that consumers always get the best price effortlessly.

Imagine you want to purchase a new electronic device. Instead of searching through different stores and comparing prices, your personal AI assistant does so in microseconds, finding the best available deal. As all companies use AI to set optimized and competitive prices, price differences between suppliers become virtually nonexistent. The marginal cost of production is so low that companies can offer products at minimal prices, focusing instead on adding value through personalization, user experience, and additional services.

In this environment, competition shifts away from price and toward innovation and quality. Companies strive to differentiate themselves not through cost but by their ability to meet consumers' specific needs and desires. This encourages a virtuous cycle of continuous improvement, driven by human creativity and the limitless potential of artificial intelligence.

The abundance generated by this new economy has profound implications for society. With basic needs guaranteed, people are free to pursue passions, develop skills, and contribute to their communities in meaningful ways. Education becomes a lifelong pursuit, not merely a preparation for employment. Arts, science, philosophy, and other fields of knowledge flourish, fueled by curiosity and the innate desire to explore and understand the world.

Cities can be redesigned to promote well-being and social interaction. Green spaces, cultural centers and communal areas replace traditional industrial and commercial zones. Mobility is enhanced by autonomous and sustainable transportation systems, reducing traffic and pollution. Artificial intelligence efficiently manages urban infrastructure, optimizing resource use and improving quality of life.

In healthcare, advances in AI and robotics enable precise diagnostics, personalized treatments, and preventive care. Diseases once considered incurable can be effectively treated or managed. Life expectancy increases, and people age with health and vitality, fully enjoying the opportunities life offers.

Globally, abundance can help mitigate conflicts caused by resource scarcity. With technologies capable of producing clean and abundant energy, sufficient food to feed the world's population, and potable water through efficient desalination, many of the challenges faced by developing countries can be overcome. International cooperation is strengthened as shared priorities of prosperity and sustainability take center stage.

However, embracing abundance also requires addressing ethical issues and ensuring that the transition to this new reality is conducted fairly and inclusively. The concentration of power and control over AI and robotics technologies could lead to significant imbalances if mechanisms are not in place to ensure that benefits are widely distributed.

The governance of artificial intelligence is an emerging field that seeks to define guidelines for the responsible development and implementation of these technologies. Principles such as transparency, accountability, and respect for privacy must be incorporated from the outset. Moreover, education plays a crucial role in preparing society to understand and interact with AI in an informed and critical way.

The relationship between humans and machines will also evolve. Rather than completely replacing human interaction, humanoid robots can complement our abilities by taking on dangerous, tedious, or strenuous tasks. This allows humans to focus on activities that require empathy, creativity, and moral judgment. Collaboration between humans and machines can result in innovative solutions to complex problems, combining the best of both worlds.

Culture and the arts may experience a new era of expression and diversity. With more time and resources, people can explore creative forms of communication and creation. Artificial intelligence can act as a tool that expands artistic possibilities, enabling new forms of music, literature, cinema, and other arts. Collaboration between human artists and AIs could lead to works that challenge and expand our understanding of aesthetics and the human experience.

In family life, abundance provides opportunities to strengthen bonds and dedicate quality time to loved ones. Children's education can be personalized, addressing individual needs and cultivating unique talents. Families can engage in community projects, travel, and activities that enrich life and promote values such as compassion, respect, and responsibility.

Spirituality and the search for meaning may also take on new dimensions. With reduced material concerns, people have more space to reflect on existential questions and explore different philosophical and religious perspectives. This could lead to greater intercultural understanding and a renewed sense of collective purpose.

In summary, embracing the abundance offered by artificial intelligence and humanoid robots is an opportunity to redefine society in more positive and human terms. While challenges exist, the possibilities for progress and improvement are immense. By focusing on collaboration, ethical innovation, and the appreciation of human potential, we can build a future where prosperity is shared, and quality of life is elevated for everyone.

The journey to this future requires vision, courage, and commitment. We must be willing to question old paradigms, explore new ideas, and work together to overcome obstacles. Technology is a powerful tool, but it is human wisdom that will determine how it is used. By embracing abundance responsibly and with hope, we can transform not only our own lives but also the destiny of humanity.